M10/Achilles

A VISUAL HISTORY OF THE U.S. ARMY'S WWII TANK DESTROYER

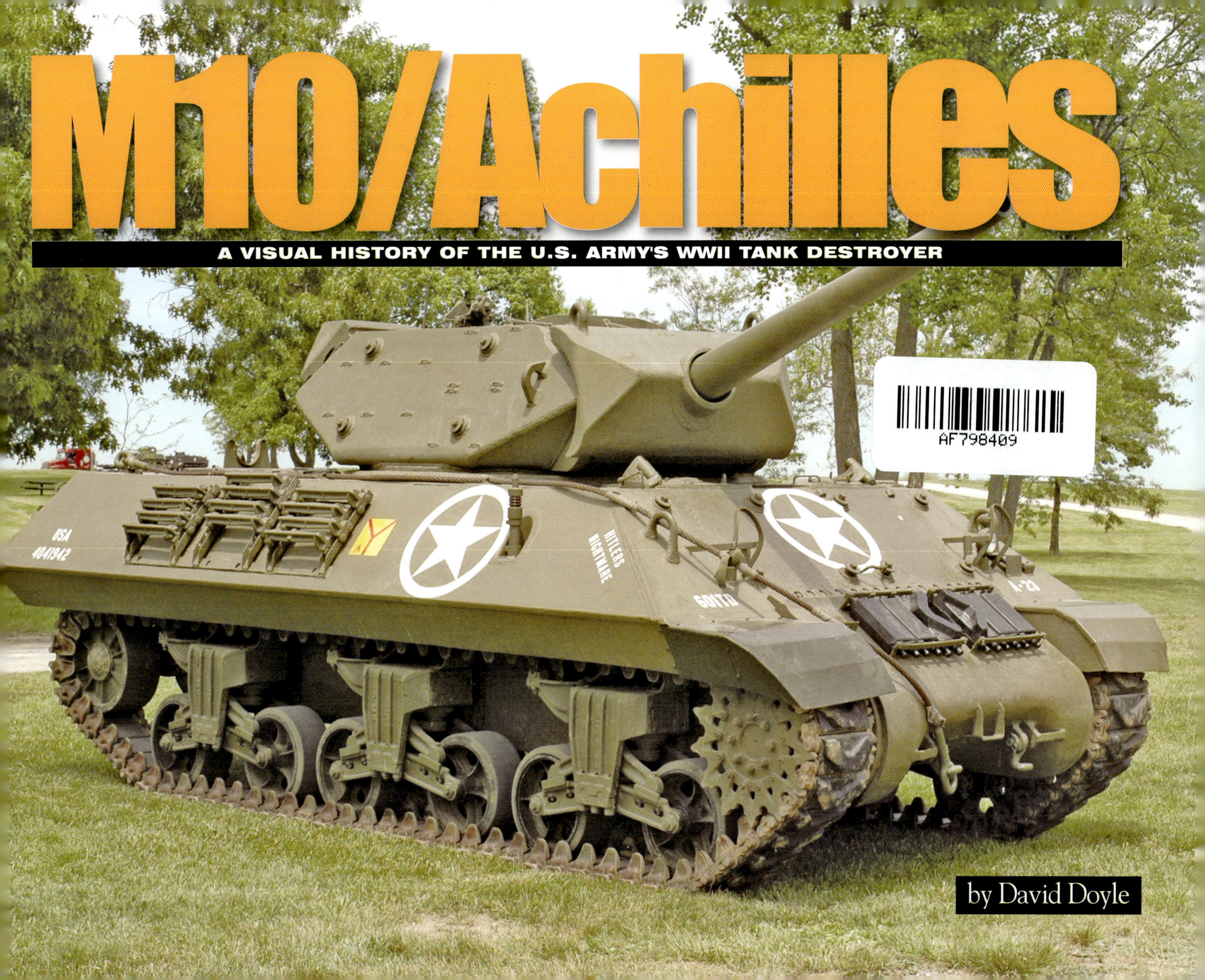

by David Doyle

Published by
The Ampersand Group, Inc.
In cooperation with HobbyLink Japan
235 NE 6th Ave., Suite B
Delray Beach, FL 33483-5543
561-266-9686 • 561-266-9786 Fax
www.ampersandpubco.com • www.hlj.com

Acknowledgements:

This book would not have been possible without the generous assistance and resources of many friends and institutions. Among them, Tom Kailbourn, Rob Ervin; Jim Gilmore; Don Moriarty; David Fletcher; Marc Sehring; Allan Cors; Joe DeMarco; Pierre-Olivier Buan; John Blackman; Chris Hughes; Mike Koening; Scott Taylor; Paul Bird; the late Jacques Littlefield and the former staff of the Patton Museum. Special thanks to my wife Denise, without whose help and support this book would not have been possible.

All unattributed photos are copyright of the author.

Sources:

Record of Army Ordnance Research and Development, 3-inch Gun Motor Carriages; Anne B. Jones, Development Visualization Branch, Ordnance Department, 1944

Report of Test of 3" Gun Motor Carriage M10, Tank Destroyer Board, 1942

Tank Destroyer History, Period inception to 15 Nov 43, Historical Officer, Tank Destroyer Center, 1944

TM 9-752A Technical Manual 3-inch Gun Motor Carriage M10, War Department, 1942

Sherman A History of the American Medium Tank, R.P. Hunnicutt, Presidio Press, 1978

Front Cover: As seen in this vehicle owned by the National Museum of Americans in Wartime, the turret, mantlet, and upper hull of the M10 were very angular in design. The final drive assembly on the bow, the lower hull, the suspension, and the tracks were borrowed from the Sherman medium tank. (Marc Sehring)

Rear Cover: The centerpiece of the M10 tank destroyer was the 3-inch gun, which packed a considerable punch and was capable of defeating many of the armored vehicles in the Axis inventories. (Marc Sehring)

Title Page: At the time it was photographed, this M10 was in the collection of the Patton Museum. It bears replica markings for Company A, 601st Tank Destroyer Battalion, and the nickname "HITLERS NIGHTMARE" on the front of the side of the upper hull. (David Doyle)

Introduction

During the war, the American strategy concerning combatting tanks was summarized by the Historical Officer of the Tank Destroyer Center as "Antitank doctrine of the United States Army at the time of the fall of France ran contrary to our historical military policy in that it implied passive defense rather than offensive action. Current field manuals stressed the importance of antitank mines and guns; but so far as their employment was taught, they were purely defensive weapons. Only the armored divisions thought in terms of aggressive methods against tanks." This final statement further hints at another detail of the planning of the day—each arm of Army was responsible for its own antitank defense.

Gradually, the War Department began to move toward a strategy built on the aggressive use of mobile antitank units employing "stealthy reconnaissance and destroy the enemy armored elements by gun fire from carefully chosen locations."

While retrospectively, another tank seems an obvious solution, but strategists of the day preferred an antitank weapon that was faster and better armed than tanks of the day—and not insignificantly, less expensive than a tank.

In fact, this last item was clear in General Leslie McNair's remarks that "Certainly it is poor economy to use a $35,000 medium tank to destroy another tank when the job can be done by a gun costing a fraction as much. Thus the friendly armored force is freed to attack a more proper target, the opposing force as a whole, in much the same manner as seacoast defenses free the Navy for defensive action at sea."

From this mindset in late 1941 the concept of a separate tank destroyer force began to jell. This force would be armed with specialized weapons. While the army's primary antitank weapons of the late 1930s were 37mm towed anti-tank guns, these were soon deemed to be inadequate against enemy armor. Accordingly, the quest for larger weapons began, as did the desire for a self-propelled antitank gun, or Gun Motor Carriage. The initial efforts which involved adapting $\frac{3}{4}$-ton Dodge trucks to mount antitank weapons (the M6, née WC-55), which were intended as interim and training vehicles pending development of the M8 armored car for use by the tank destroyer reconnaissance force. In the same manner, half-tracks were fitted with cannon, yielding the interim M3 75mm Gun Motor Carriage. But the quest was on to create a specialized and ideal tank destroyer, utilizing a tank chassis as the basis.

Two failed attempts were made to mount the 3-inch gun in a modified hull of an M3 medium tank. By January of 1942 a prototype was in the works to mount the weapon in an open-topped turret on the chassis of the twin GM Diesel-powered M4A2 Sherman Medium tank. After some months of development a design was finalized for a vehicle sharing the suspension, lower hull, and engine with the M4A2 but with an upper hull made up from thinner, but sloping, armored plate. It was hoped that the weight savings would produce increased automotive performance. Initially designated T35E1, when the design was standardized it was redesignated M10. In addition to the 6,700-plus Diesel-powered M10 tank destroyers, a further 1,700 M10A1 vehicles were built, these being driven by Ford GAA gasoline engines.

While at the time they were fielded many felt that the M10 and M10A1 were the best self-propelled antitank weapons on the battlefield, they failed to impress General Andrew D. Bruce, the first head of the Tank Destroyer force. The M10 was standardized despite his objections, however he continued to view the vehicle, like the M3 Gun Motor Carriage, as an interim vehicle.

Bruce was not alone in feeling that while the 3-inch weapon of the M10 was superior to that found on earlier U.S. tank destroyers, it was inadequate against the ever-increasing weight of German armor. The British addressed this by rearming some of the 1,700 M10s that they received with the superb Ordnance Quick Firing 17-pounder antitank gun. From May 1944 through April 1945, 1,017 of these vehicles were converted at various Royal Ordnance Factories. These vehicles were designated by the British as 17-pdr. SP M10 Mark 1c. After the war the name Achilles was given these vehicles.

The U.S., also seeking heavier armament, created a new turret, mounting a 90mm gun. Beginning in April 1944 the new 90mm-armed tank destroyer, designated M36, entered production. These vehicles were built new from 100 new hulls purpose-built for this, and by remanufacturing M10A1s—primarily from U.S.-based training units. As the supply of these chassis was depleted, converting Diesel-powered M10s, resulting in the M36B2, created additional vehicles.

While U.S. antitank doctrine changed, rendering all the tank destroyers obsolete post-WWII, many of these vehicles were supplied to other nations, and in fact some survived as combat vehicles into the 21st century.

One of the prototypes for the M10 tank destroyer was this vehicle, the 3-inch Gun Motor Carriage T35, assigned U.S. Army serial number 4027647. Based on the M4A2 medium-tank chassis, the T35 featured a rounded, cast turret with a bulge around its vertical center, mounting a 3-inch Gun T12, later standardized as the 3-inch Gun M7. The T35 is shown at Aberdeen Proving Ground on 23 April 1942. (NARA)

The turret of the T35 had only a partial roof with a fixed periscope mount for the gunner on the right side and a rotating periscope on the left side. The interior of the turret was very spare and was painted white, as were the guards for the 3-inch Gun M7. The turret could traverse 360 degrees, and the gun had a range of elevation of -10 to +30 degrees. (NARA)

A second prototype for the M10 tank destroyer was the 3-inch Gun Motor Carriage T35E1, shown in a 4 May 1942, Aberdeen Proving Ground photograph. Its lower hull, turret, and main gun were the same as the T35's, but the upper hull featured sharply angled armor plates, 1-inch thick, with inward-sloping ½-inch plates at the bottom forming a sort of skirt to protect the tops of the tracks and the lower hull. (NARA)

The T35E1 is viewed from the right side. During the tests of the vehicle at Aberdeen in the spring of 1942, certain modifications were made. For example, the ratio of the elevating gear was reduced from 570:1 to 1,700:1, and the manual traversing gear was reduced from 78:1 to 400:1. In firing tests against both prototypes with .30-caliber and .50-caliber ammunition, the T35E1 fared better than the T35 because of its sloping armor. (NARA)

Like the sides of the upper hull of the T35E1, the rear of the upper hull incorporated an angled main plate, with an inward-angled narrower plate below it. Two lifting lugs were welded to the upper plate. Faintly visible below the lower plate are the resonance chambers and the exhaust tailpipes. (NARA)

The T35E1 was standardized as the M10 in June 1942, but the M10 had a much different turret, pentagonal rather than rounded in plan, and fabricated with welded, angled plates. Designated the 3-inch Gun Motor Carriage M10, it went into production at the Fisher Tank Arsenal, Grand Blanc, Michigan, in September 1942. Several examples of early M10s are seen here under construction at Fisher. (Library of Congress)

M10 tank destroyers are being tested for their performance in deep mud at Fisher Body Division's proving ground at Grand Blanc in March 1943. (University of Memphis)

Top left: A gun crew is in the turret of an early-production M10. Grousers are stowed on the rear of the turret: an early move to provide some counterweights to offset the heavy weight of the gun placed on the front of the turret. Tests of the stock turret showed that even a slight sideward cant of 3.6 degrees made traversing the turret, and thus tracking a moving target, much more difficult because of the weight the barrel imposed on the front of the turret. The .50-caliber machine gun mounted on the rear of the turret had the added benefit of placing more weight on that part of the turret. (Patton Museum) **Top right:** An early M10 is seen with grouser storage racks on the rear of the turret. The round, raised features on the side of the hull and the turret were mounting bosses for applying supplemental armor. The bosses were welded to the armor and were tapped with threaded holes for bolts and washers to secure the supplemental armor. The bosses had a thickness of ¾ inch, allowing for spacing between the main armor and the supplemental plates. However, this armor was never applied to the M10s operationally. (Patton Museum) **Above left:** Tests of the early-production 3-inch Gun Motor Carriage M10 by the Tank Destroyer Board in late 1942 determined that the vehicle provided a firing platform that was extremely stable at all firing positions. The firing results of the 3-inch gun were considered satisfactory when firing at stable or moving targets but much less so when the vehicle was positioned on a slope. The advent of proper counterweights for the turret would solve this problem. (Patton Museum) **Above right:** By the time the M10 tank destroyer made its combat debut in North Africa in March 1943, the turrets were fitted with new, slab-shaped counterweights, two per turret, with a space between them. Sometimes referred to as "quick-fix" counterweights, they are visible here on two M10s, turrets traversed to the rear, at a small settlement in North Africa. With a combined weight of 2,400 pounds, these counterweights significantly improved the traversing of the turret, but improved counterweights lay ahead.

Because of the high priority the army assigned to M10 production, Ford was selected to manufacture a version designated the M10A1. Based on the Ford-built M4A3 medium tank chassis, it was powered by the Ford GAA V-8 gasoline engine, which replaced the GM 6046 Diesel engines of the M10. Seen here are M10A1 turrets undergoing assembly at Ford's Highland Park, Michigan, factory. Ford's system was to arrange the turret plates on a former ring, tack-weld the plates, and then transfer the assembly to rotating and tilting fixtures, as seen to the right of the photo, for ease of final welding.

Welders at the Ford Highland Park plant perform final welding on the joints of the turret plates for M10A1 tank destroyers. The assemblies are on the special tilting and rotating fixtures that Ford employed.

Assembly is progressing on a line of M10A1 tank destroyers at Ford's Highland Park facility. To the left, a turret is installed on a hull, with the gun and mount still to be installed. To the right is a stand for M4A3 medium-tank turrets; assembly of those tanks was continuing on a parallel assembly line. In the background, the triangular shapes on the M10A1 chassis are the baffles on the insides of the engine-compartment doors (or grilles).

Ford assembly-line workers go about their tasks, with M10A1s on the left line and M4A3 medium tanks to the right. These turrets are equipped with wedge-shaped counterweights, which superseded the slab-shaped "quick fix" counterweights. Designed by Fisher Tank Arsenal, these counterweights were assigned part numbers E7992 (right) and E7993 (left) and weighed a combined 3,700 pounds. On the floor between the two assembly lines is a roll of tracks and, farther away, several M10A1 turret assemblies.

A completed 3-inch Gun Motor Carriage M10A1 stands on the factory floor at Highland Park. Webbing straps are attached to footman loops on the turret for stowing bed rolls. Between the two front bosses on the side of the hull is a recess for a radio antenna base. Above the rear of the turret is a mount for an antiaircraft machine gun.

Once completed, the vehicles were hoisted aboard rail cars, affording this clear view of the bow, the belly, and the steel T49 tracks of an M10A1 at the Ford factory. The armored casting at the front of the hull is the final drive assembly (commonly referred to as the FDA). The foundry mark "SS" in an oval toward the bottom of this casting stands for the Scullin Steel Company of St. Louis, Missouri. A crew step is welded to the lower front of the final drive assembly.

A newly finished M10A1 tank destroyer with wedge counterweights bears U.S. Army registration number 4047125 toward the rear of the side of the hull. By now, some of the supplemental-armor bosses on the sides of the hull were being put to use holding storage racks for grousers: steel fixtures that were attached to the tracks to give them added traction on snow, ice, and soft ground.

General Data

Model	M10	M10A1
Weight*	65,200 lbs.	64,000 lbs.
Length	286.3 inches	286.3 inches
Width	120 inches	120 inches
Height	114 inches	114 inches
Track	83 inches	83 inches
Crew	5	5
Max Speed	25 mph	26 mph
Fuel Capacity	165 gallons	168 gallons
Range	200 miles	115 miles
Electrical	24-volt negative ground	24-volt negative ground
Transmission Speeds	5	5
Turning Radius	31 feet	31 feet
Armament main	3-inch gun M7	3-inch gun M7
Flexible	1x .50 M2HB	1x .50 M2HB

Overall dimensions listed in inches. Measure with main gun facing forward, and an anti-aircraft machine gun mounted.
*Combat weight

Engine Data M10

Engine Make/Model*	GMC 6046 or 6046D Diesel
Number of Cylinders	12
Cubic Inch Displacement	850
Horsepower	375 @2,100
Torque	855 @1,300
Governed Speed (rpm)	2,100

Engine Data M10A1

Engine Make/Model*	Ford GAA
Number of Cylinders	60 Degree V-8
Cubic Inch Displacement	1,100
Horsepower	450 @2,600
Torque	950 @ 2100
Governed Speed (rpm)	2,600

Radio Equipment

The M10 was equipped with SCR 610 radio set, RC99 interphone, and a M238 flag set.

Completed M10A1s are rolling out of Ford's Highland Park factory. All of these vehicles had wedge counterweights on the turrets and bosses for mounting supplemental armor: a feature that in July 1943 would be eliminated as a production feature except for the glacis plate on the M10s and M10A1s. In the background to the right are two completed M4A3 medium tanks. (University of Memphis)

Registration Numbers

Model	Quantity	Contract Number	Registration Numbers	Serial Numbers
M10	1,800	374-ORD-1880	4040705 - 4042504	3 - 1802
M10	1,200	374-ORD-1880	4081054 - 4082253	2839 - 4038
M10	1,117	374-ORD-1880	40110110 - 40111226	5991 - 7107
M10A1	1,038	374-ORD-1213	4046509 - 4047546[1]	1803 - 2840
M10A1	375	374-ORD-1213	40112380 - 40112754	7984 - 8358
M10/A1	876	374-ORD-1880	Unrecorded[2]	Unrecorded

1. 200 registration numbers cancelled. **2.** Information derived from "Armored, Tank, and Combat Vehicles 1940-1945." For reasons unknown, the 876 vehicles on Fisher's contract W-374-ORD-1880, Production Order Number T-7581 were omitted from the original document. Notably, this order also contained as many as 300 M10A1s.

A 1942-production, Fisher-manufactured M10 tank destroyer, U.S.A. number 4041128, is observed from overhead. The nickname "Deadeye" is painted on the hull. No counterweights are present. A Browning M2 .50-caliber machine gun is on a pintle mount at the rear of the turret. Swivel-mounted rods, three on the top front of the turret and four at the rear, were designed to support a canvas top. (Patton Museum)

This close-up photo of an M10 of the 803rd Tank Destroyer Battalion at Bodmin, Cornwall, England, on 7 April 1944, shows the fourth and final design of counterweight, the "duckbill." Introduced in June 1943, it had a flat top, a scooped underside, and built-in pockets for stowing gear. Along with the duckbill counterweight came a new, enlarged turret, on which the two upper plates at the rear of the turret were nearly vertical, making for more room to the rear. Note the .50-caliber machine gun with a canvas spent-brass collection bag on the right side, and the three ready rounds of 3-inch ammunition in the rear of the turret. To the front of the canvas breech cover for the 3-inch gun is the interior travel lock, to the right of which is a panoramic sight, for use in indirect fire. For direct fire, the gunner used a telescopic sight. (Patton Museum)

The breech of the 3-inch gun in an M10 turret is the focus of this photograph taken at a vehicle depot. On the right side of the breech is the hand-cocking lever; below the lever is the firing solenoid. To the left of the interior travel lock above the gun is the elevation rod, which is attached to the turret roof. The yoke-shaped tube to the front of the breech is the hand-firing lever. In the left front corner of the turret is the gunner's telescopic sight and head guard. To the far right are the panoramic sight and its bracket. (Patton Museum)

Power for the M10A1 tank destroyers was provided by the Ford Tank Engine Model GAA. This was a gasoline-fueled, V-8, four-cycle design with an 1,100 cubic inch displacement, a net horsepower of 450 at 2,600 rpm, and a gross horsepower of 500 at 2,600 rpm. This engine later would be used in the M36 tank destroyer. (Patton Museum)

The engine used in the M10 tank destroyers was the General Motors 6046, a Diesel-fueled twin in-line, 12-cylinder design, comprising two side-by-side GM 6-71 truck engines, with six cylinders per engine block. An advantage of this engine was that if one of the engine blocks became disabled, the other block could still drive the vehicle, although at reduced power. (Patton Museum)

A G.I. hands off a 3-inch round to a crewman in the turret of an M10 tank destroyer with wedge-type counterweights. Below the rear overhang of the hull is a scoop-shaped exhaust deflector. A holder for five-gallon liquid containers has been improvised with an angle iron attached to long, threaded rods fastened to two of the supplemental-armor bosses. (Patton Museum)

This is one of the M10s that served in the fighting in Tunisia in early 1943, and it was photographed on 13 March. Although most of the markings on the side of the hull are lost in the glare, there was a diamond with the number 1 in it toward the front, and the nickname "INVINCIBLE" in a rectangle farther to the rear. (NARA)

A dust-caked M10 tank destroyer of Company A, 899th Tank Destroyer Battalion, leads a column along a road in the desert outside of Maknassy, Tunisia, on 8 April 1943. The vehicle is running on tracks with plain rubber track shoes. (Patton Museum)

Seen from the rear is an M10 at the Ordnance Division, Mediterranean Base Section, Oran, Algeria, on 5 April 1943, with a good view of the shapes of the wedge-type counterweights on the turret. A cover is placed over the antenna base in the recess on the forward end of the side of the hull, and a folded tarpaulin is strapped to the side of the turret. (NARA)

A U.S. Army M10 advances to the front during a fight at Bir Marbott Pass in the El Guettar area of central Tunisia in April 1943. The turret has the quick-fix, slab-type counterweights: two knapsacks are hanging on the left one. The vehicle has been given a mottled camouflage scheme. (Image Bank WW2)

Two French Army crewmen of an M10 are on the lookout to the rear during training maneuvers in North Africa. The nickname "AMIENS" is painted on the mantlet. Grousers are secured to the rack on the side of the hull, and a section of spare track is lying on the rear deck. The counterweights on the rear of the turret appear to be the quick-fix, slab type. (Image Bank WW2)

A U.S. Army M10 tank destroyer is proceeding to the front in the Cassino Sector of Italy on 6 November 1944. The white identification stars on the hull, the glacis, and the turret counterweights have thin, probably yellow, circles around them. (Patton Museum)

New Guineans in military khakis are given rides on U.S. Army M10s at the ordnance motor pool in Cavendish, Brisbane, Australia, on 24 November 1943. The vehicle to the left is named "Yehudi." Stencils on the M10 to the right concern tow cable stowage. (NARA)

During the Battle of Kwajalein in the Marshall Islands, 31 January -4 February 1944, five M10 tank destroyers were assigned to the Assault Gun Platoon, Headquarters Company, 767th Tank Battalion. One of those vehicles is seen here during the fighting on that atoll. Note the gun travel brace on the rear of the deck, on which the gun barrel would rest during transit. (Patton Museum)

Two of the M10 tank destroyers assigned to the Assault Gun Platoon, Headquarters Company, 767th Tank Battalion, were photographed together during the Battle of Kwajalein. Note the teardrop-shaped, swiveling cover for the gunner's sight aperture on the mantlet. (Patton Museum)

In a view of three M10 tank destroyers resting in a forest, all three have canvas covers erected over the turrets. The vehicle to the far left has slab-type counterweights, while the one to the right has wedge counterweights. Note how the wedge counterweights stand off from the rear plate of the turret. The 3-inch gun of each M10 is resting in the gun travel brace. (NARA)

An M10 negotiates a narrow lane in a town near La Manche, France, on 7 July 1944. On the right mudguard is a small reel of communications wire. The unit markings on the right side of the final drive assembly has been censored; on the left side of the FDA are markings for Company B and the 21st vehicle in the line of march. (NARA)

An M10, with another one trailing behind it, advances down a rural lane near Saint-Lô on 24 June 1944. Visible in small characters on the lower left corner of the glacis are markings for the 899th Tank Destroyer Battalion, First Army. The nearest M10 is running on T48 tracks, with rubber-clad track shoes with chevron treads. (NARA)

The blast of the main gun has kicked up leaves on the pavement as an M10 engages German snipers in the tower dimly visible in the distance, in Aachen, Germany, on 15 October 1944. A marking for the First Army is next to the left taillight, but the unit markings below it have been censored. This vehicle was equipped with T51 tracks. (NARA)

Top left: The British Army employed M10 tank destroyers, designating them Wolverines. This example, which has come ashore in Normandy on D-day, has a canvas mantlet cover; T48 rubber tracks with chevron treads; two bogie wheels on special holders with Hessian-tape camouflage netting over them; and a .50-caliber machine gun atop the turret. It is also fitted with the early version of the E8543 final drive assembly. **Above left:** Despite their principal role as tank destroyers, M10s sometimes were diverted to fire-support missions. This British M10 Wolverine of A Troop, 111th Battery, 2nd Anti-Tank Regiment, 6th Armoured Division, has been emplaced on an earthen ramp to increase the gun's maximum elevation and is firing on German positions near the Arno River outside of Firenze (Florence), Italy, in August 1944. (Patton Museum) **Above right:** An M10 with deep-fording kit installed is being lowered from a freighter at a French port, to join three similarly appointed M10s parked on the deck of a landing craft below. These vehicles were serving with Free French forces. (Patton Museum)

U.S. Army Ordnance personnel are removing the 3-inch gun of an M10 for repair or replacement with the help of a M1A1 wrecker boom at a base in Normandy. Removing the gun was not a major operation and was accomplished by removing the telescopic sight, the elevating mechanism, and the trunnion caps, and then lifting the gun mount off the trunnions and pulling it out of the turret. (Image Bank WW2)

Troops of the 3rd Battalion, 22nd Infantry Regiment, hitch a ride on an M10 tank destroyer still wearing an adapter for deep-water fording exhaust trunk, outside of Mabompré, Belgium, in early September 1944. The vehicle is running on well-worn rubber-chevron tracks. On the upper rear plate of the hull is a stowage rack fabricated of steel strapping. (Jim Gilmore collection)

M10s assigned to A Company, 634th Tank Destroyer Battalion, pick their way through a narrow lane in the German city of Aachen, located on the Siegfried Line on the western border of Germany, on 14 October 1944. These vehicles were on a mission to destroy several German observation posts in the city. To supplement the 1.5-inch rolled armor of the glacis, the crews have arranged sandbags on the glacis plates. The nearest M10 is missing its right fender. (NARA)

The Battle of Aachen was a bloody, hard-fought urban fight that lasted almost three weeks in October 1944. Here, an M10 assigned to the 634th Tank Destroyer Battalion, one of several TD battalions to participate in the battle, fires a 3-inch round over a fence at German defenders in that city. This M10 was the number 14, as indicated by a marking on the right side of the rear of the hull: in the background is another M10, number 13. (NARA)

Two widely spaced M10s assigned to Company C, 893rd Tank Destroyer Battalion, proceed along a gloomy stretch of road in the Hürtgen Forest on 4 November 1944, en route to Schmidt, Germany, to contend with a German tank force spotted there. Visible on the bow of the front M10 is some type of netting or wire mesh, for holding tree branches for camouflage purposes. (NARA)

Several M10s of the 1st Battalion, 11th Antitank Regiment, 6th South African Armoured Division, are firing at German forces flanking the Fifth Army in the mountains along Highway 64 near Bologna, Italy, on 17 November 1944. On the turret counterweight of the nearest M10 is the nickname "VAGABOND," while the next one is nicknamed "VENTURE." (NARA)

Four M10s of the same unit shown in the preceding photo, the 11th Antitank Regiment, 6th South African Armoured Division, are seen from another perspective. Tarpaulins are at the ready to protect the interiors of the vehicles from moisture and cold. Note how the nearest M10 is resting in a prepared, leveled position, which would enhance the accuracy of its 3-inch gun while engaged in indirect fire. (NARA)

An M10 picks its way through a village in the European Theater in the final months of World War II. Chicken wire has been fastened to the sides of the turret and hull for holding tree branches: so-called "local camouflage." A makeshift storage rack on the rear of the hull holds a motley collection: a funnel, knapsack, tarpaulin, wire, and other material. The counterweight is the duckbill style, and a cover is erected over the turret. (NARA)

As part of the 35th Infantry Division, the 654th Tank Destroyer Battalion supported the 134th Infantry Regiment in taking Habkirchen, Germany, in mid-December 1944. In this photo, an M10 of that battalion is partially exposed at an intersection in that town on the 15th of the month, during a multi-day mopping up operation. Sandbags are piled on the glacis for extra protection, and the barrels of a .30-caliber machine gun and a .50-caliber are visible above the turret. (NARA)

Along Ormoc Beach, Leyte, in the Philippines, the crew of an M10 with the 632nd Tank Destroyer Battalion prepares the vehicle for further operations on 16 December 1944. Two of the men are working on something under the deep-water fording exhaust trunk. An M1 rifle and a webbing belt with canteen hang from the 3-inch gun barrel. The .50-caliber machine gun on the turret has its dust and weather cover installed. A Japanese barge is in flames in the background. (NARA)

An armored-enclosure kit installed on an M10 turret with duckbill counterweights is seen from the left side. The front and rear shields were welded to the top of the turret. The front and rear roof panels were welded to the tops of the shields in a manner such that the edges of the roof panels were visible. Two hatch-door panels were hinged to the front roof panel. Vision slots were cut into the shield for all-around vision. **Inset:** A cover fabricated from 5/8-inch armor plate has been installed over the forward half of the turret opening of an M10 assigned to the 813th Tank Destroyer Battalion in the Soultz area of France in mid-January 1945. The plates are heavily pitted and apparently were salvaged from scraps of armor. Two rectangular vision slots have been cut through the frontal plate only. (NARA, both)

An M10 tank destroyer advances east along Grand Rue at the center of Rohrwiller, France, on 4 February 1945. The taking of Rohrwiller was a 36th Division operation, so this vehicle would have been attached to the 636th Tank Destroyer Battalion, that division's TD battalion at that time. In the background is the shell-pocked Église Saint-Wendelin, which still stands. (NARA)

A German strongpoint along the Sauer River in Belgium is the target of the moment for an M10 assigned to the 5th Infantry Division, Third Army, on 7 February 1945. At that time, two TD battalions were detailed to that division: the 737th and the 803rd. Liquid containers, rations boxes, bedrolls, and camouflage netting are stashed in two racks on the rear of the hull. (NARA)

A column of M10s of the 644th Tank Destroyer Battalion, 8th Infantry Division, rolls through Düren, Germany, on 24 February 1945. The nearest M10 (and likely the second one) is equipped with a hedgerow cutter, a device that allowed the vehicle to crash through thick brush and hedgerows. Because it was often difficult and dangerous to bring the 50-caliber machine gun into play against threats to the front, some tank destroyer units started mounting .30-caliber machine guns on the fronts of the turrets by early 1945, and one such weapon is seen here. (NARA)

he rear of an M10 tank destroyer presents itself as it advances toward Jülich, Germany, on 25 February 1945. Crew knapsacks are arranged around the rear of the duckbill counterweight. Faintly visible in the distance between this vehicle and the soldier walking alongside it is another M10. (NARA)

Forces of the 1st Infantry Division are poised for the advance to the Rhine River in Gladsbach, Germany, on 1 March 1945. On the side of the street are two M10 tank destroyers, with several Sherman tanks to the rear. The M10s probably were attached to the 634th Tank Destroyer Battalion, which, according to the 1st Division's order of battle, was the only TD battalion assigned to the division at that time. (NARA)

The M10 tank destroyer to the right and another one in the distance are entering Neuenhausen, Germany, during the 1st Infantry Division's advance on 3 March 1945. Here again, these vehicles likely were serving with the 634th Tank Destroyer Battalion. Both vehicles have wedge counterweights. (NARA)

In the rubble of Sankt Andreasberg, Germany, an M10 tank destroyer, probably with the 634th Tank Destroyer Battalion, stands ready to deal with any remaining enemy resistance in that town on 15 April 1945. A .50-caliber machine gun is mounted toward the front of the turret. (NARA)

At the front of a group of Allied vehicles entering Bologna, Italy, on 21 April 1945, is an M10 tank destroyer with a sergeant wearing a bush hat on the turret roof. The scout cars and carrier in the photo are British types, but the headgear seen on the crews and visible insignia indicate the 6th South African Division. (NARA)

A contingent of freshly washed and neatly arranged M10s of the South African 6th Armoured Division pass the reviewing stand during a victory parade at the Autodrome at Monza, Italy in May 1945. Tarpaulins cover any stowed gear on the rear decks. (NARA)

This 3-inch Gun Motor Carriage M10, owned and restored by the National Museum of Americans in Wartime, features the mid-production turret with wedge-shaped counterweights. Casting numbers are visible on the upper right of the final drive assembly, sometimes called the differential cover. A siren is on a bracket to the left side of the left headlight assembly. (Marc Sehring)

Mix-and-match bogie wheels sometimes were encountered on M10s during World War II, as crews replaced damaged or worn-out units with whatever wheels were available. This M10 has three different types of bogie wheels on the left side. From the front, the first, second, fifth, and sixth wheels are solid-disc concave; the third wheel is solid-disc convex, and the fourth one is solid-spoke. (Marc Sehring)

Top left: This restored M10 tank destroyer bears the name "Hitler's Nightmare" at the front of the side of the hull. This vehicle lacks the bosses for mounting supplemental armor on the sides of the hull, a feature discontinued at the assembly lines in July 1943. However, the bosses remained on the sides of the turret and on the glacis. (Paul Bird) **Top right:** The M10 was designed with sloped armor plate on the upper hull to maximize its protection. The glacis armor is $1\frac{1}{2}$-inches at 55 degrees, the sides $\frac{3}{4}$-inch at 38 degrees, significantly thinner than the M4 series. (Don Moriarty) **Above left:** The left side glacis plate mounts the left front lift ring, headlight guard, and a removable combat headlight, absent here. The lift rings are cast integral with their base and welded to the glacis armor. **Above right:** The M10 was built on the lower hull and running gear common to the mid-production Diesel-powered M4A2 Sherman. The left side bogie mounts the stamped face road wheel and straight return roller bracket.

An M10 from the post-July 1943 production of the vehicle is viewed from the rear. Bosses for supplemental armor are not present on the side of the hull. The general arrangement of the upper plate of the rear of the hull is illustrated. An exterior telephone box for the use of troops outside of the vehicle needing to communicate with the crew (a post-war modification), is on the right side. (Paul Bird)

An M10 with the supplemental-armor bosses and wedge-type counterweights on the rear of the turret is seen from the rear. From the rear the inclination of the various plates of the upper hull are evident. The rear plates of the upper hull and lower skirt are inclined at 38 degrees. The upper plate is $^3/_4$-inch thick, the lower skirt $^1/_4$. The part number of the right counterweight, E7992, is on the side of the counterweight, in raised letters on a slightly raised rectangle. Next to the part number are the raised letters TA, applied upside down with reference to the part number. This vehicle is part of the collection of the French tank museum at Saumur.

The Saumur vehicle is seen from the rear with all of the pioneer tools missing from their sheet metal brackets. The lower skirt is stained with exhaust soot from the twin General Motors Diesel engines. The round projections on the hull and turret sides are supplemental armor bosses. They were installed on the M10 to mount appliqué armor plate kits that were subsequently never produced.

Top left: The left rear corner of the upper hull displays the left rear taillight and hull lift ring. The taillight has a red lens at the top and a thin slit at the bottom used while operating the vehicle in blacked out conditions. **Top right:** The taillight is a standard fitting on all U.S.-built vehicles during WWII. The angled mounting bracket is specifically built to match the angle of the rear hull of the M10. The guard is built from $^1/_4$-inch mild steel. **Above left:** Each of the vehicle's pioneer tools has its own mounting bracket and is held in place by a web strap through an accompanying footman loop. The shovel and sledgehammer are mounted in the center of the rear plate. **Above right:** The fitting of the lifting ring and taillight on the right rear are mirror images of the left. The lift ring is a cast steel part with large weld beads attaching it to the upper plate. The part's drawing number is cast into the ring. (Don Moriarty, all)

Top left: Many U.S. tanks and tank destroyers were equipped with an infantry phone during depot overhauls or through Modification Work Orders (MWO), although this was not a feature seen on M10s in WWII. This BC 1362 Infantry Interphone box is tied into the vehicle's intercom system to allow communications with supporting troops. **Top right:** The right side taillight has a red brake light lens at the bottom and the blackout lens at the top. These lamps aided drivers in keeping proper vehicle spacing while driving in columns during blacked out conditions. **Above left:** The right rear corner of the upper hull shows the construction of the rear plate overlapping the side armor plate and the corresponding weld seams. The mattock head is held in its bracket by a length of web strap. **Above right:** The left rear corner welds indicate the difference in thickness between the upper plate and the lower 1/3-inch thick armored skirt. The skirt has a cutout in the rear to clear the large central exhaust deflector. (Don Moriarty, all)

Top left: The large flow of air from the two engine-mounted cooling fans and the engine exhaust required the installation of a sheet metal deflector to prevent dust and debris being churned up while driving. **Top right:** The deflector is a large sheet metal device and is similar to that used on the M4A2 Sherman. It has pivots at each side that allow it to be rotated up, providing access to the exhaust and radiators during maintenance. **Above left:** The engine exhaust pipes exit through holes in the center of the lower hull rear plate. The pipes attach to a cast exhaust collector that diverts the gases into sheet metal baffled mufflers missing here. (Don Moriarty, three) **Above right:** At the bottom edge of the lower hull rear plate is a mounting for a tow pintle flanked by two tow lugs for attaching a tow cable. Note the lower flange and bolts of the cast exhaust deflector just above the pintle mount.

Top left: The right side idler wheel mount is bolted to the rear corner of the lower hull. An eccentric axle is used to adjust track tension by loosening the three lower bolts and turning the large hex head nut at the left. **Top right:** A large spring is mounted within the cast cylinder of the tow pintle mount. This reduces the amount of bumping fore and aft or "trailer hitching" of a loaded trailer when starting and stopping the vehicle. **Above left:** The left side idler wheel mount is a mirror image of that on the right. The tension adjustment is set with a large spanner wrench stowed on the rear lower skirt. Properly adjusted the tracks will have virtually no sag. **Above right:** The idler wheel on this M10 is made from a stamped face welded to a cast steel tire portion and hub. The hub has a grease fitting and relief valve for lubrication. The reinforcing webs are a common post-WWII addition. (Don Moriarty, all)

This M10 leaves deep tracks in fairly dry ground. The relatively high ground pressure of M4-based vehicles with the original narrow track plagued the series. Note the open fuel filler cap and engine deck door. (Mike Koening)

The turret of this M10 is traversed to the 4:00 o'clock position. Note the weld seam at the lower edge of the hull side plate and the tow cable stowed along the plate's top edge. The foreground vehicle is an M4A1 Sherman. (Mike Koening)

Top left: Several types of road wheels were used on M4 series vehicles. This open spoke wheel is made up from numerous cast and flame-cut components requiring numerous man hours to assemble on a jig. The tire is pressed on. **Top right:** The inside faces of the VVSS track is smooth to contact the rubber tire portion of the road wheel and to run smoothly over the steel track return rollers and idlers. The side armor skirt is supported by welded strips of steel. **Above left:** As on all M4 series vehicles the tracks are driven by two drive sprockets at the front of the vehicle. The sprockets consist of two 13-toothed plates bolted to a cast central hub. These plates are the forged type. **Above right:** The right hull side armor plate is also fitted with the grouser rack. The grousers have one end that is bolted to the grouser casting to allow them to break down to slip over the track pins inside the track end connectors. (Don Moriarty, all)

A mid-production turret with wedge-type counterweights is mounted on this M10. Note the recesses on the front of the counterweight. Black-painted grousers are in the racks on the hull. All of the bogie wheels are of the concave, solid-disc type. (Marc Sehring)

Top left: The turret seen in almost pure profile gives a good view of the construction of the front plates behind the gun shield and the stand-off mounting of the turret counterweights. The road wheels are a post-war design. **Top right:** The tracks on this vehicle are the T48 rubber chevron style and are well-worn and packed with earth. The front fenders have been replaced with the standard fenders from a welded-hull M4 Sherman. **Above left:** At the right front of the hull is a recessed cutout for the antenna mount. The radio was mounted in the right front sponson area and was manned by the co-driver. The antenna base is a MP-48 with grounding strap fittings. **Above right:** Viewed from above the construction of the socket is evident with the casting welded into the hull side plate and the floor of the socket welded in. The hole in the side of the antenna base is for a missing grounding strap. (Don Moriarty)

The faceted design of the fronts of the mud guards of the M10 is apparent on this restored, operable M10. Two different types of rubber track links are on the two holders on the final drive assembly: four T51s with flat treads and two T48s with chevron treads. (Mike Koening)

From this angle, the forward-tilted hinges on the driver's and assistant driver's hatch doors of the Saumur M10 tank destroyer are in view. On the upper facet of the right side of the gun shield or mantlet is a casting mark; it is the part number E6475, with a five-pointed star after it (indicating production at the foundry of Symington-Gould), and the number 917 above it; these marks are upside down when viewed from the front. This vehicle mounts the later E8543 final drive assembly. These probably were first fitted to the M10 at Fisher in November 1943. It is not known if this is orignal to the vehicle, or a replacement.

Like the vehicle on the previous page, this M10 is fitted with the single piece cast E8543 final drive assembly. It can be identified by the double tow lugs on each side and the welded-on steps. (Paul Bird)

Top left: The right side glacis mounts a headlight and guard. The standard headlight was topped by a blackout lamp and was removable for safe storage. The headlight's electrical connection feeds through the glacis plate. **Top right:** The headlight guards are equipped with tubes that store a pin used to fill the headlight mounting hole when the light assembly was removed. The pin, shown in place here, kept moisture from entering the vehicle. **Above left:** The T48 rubber chevron tracks on this vehicle show characteristic wear and chipping of the blocks. The rubber blocks were cast and vulcanized around a skeletal steel framework that surrounded the track pins. **Above right:** The later style of final drive assembly is the E8543 and may not be orignal to this M10. Outboard of the tow lugs are the welded crew steps. (Don Moriarty, three)

Top left: The bolts at the upper edge of the differential cover are protected from small arms fire by a cast-in bullet splashguard. The track stowage racks indicate this being one of 1648 M10s that were supplied to Britain in WWII. **Top right:** The T48 track was developed to provide better traction in soft turf than the earlier flat rubber block T41 style of the M3 Medium. Rubber shortages brought about the development of several styles of all-steel tracks. **Above left:** The left side glacis plate is home to a headlight and guard. The guard is equipped with the empty tube for the headlight plug. The remains of the securing chain for the plug is present. Above, the driver's hatch is open. **Above right:** The two appliqué armor mounting bosses at the lower center of the glacis plate mount a length of steel rod for a crew step. Climbing the front slope of the M10 to enter the vehicle can be a difficult task. (Don Moriarty, all)

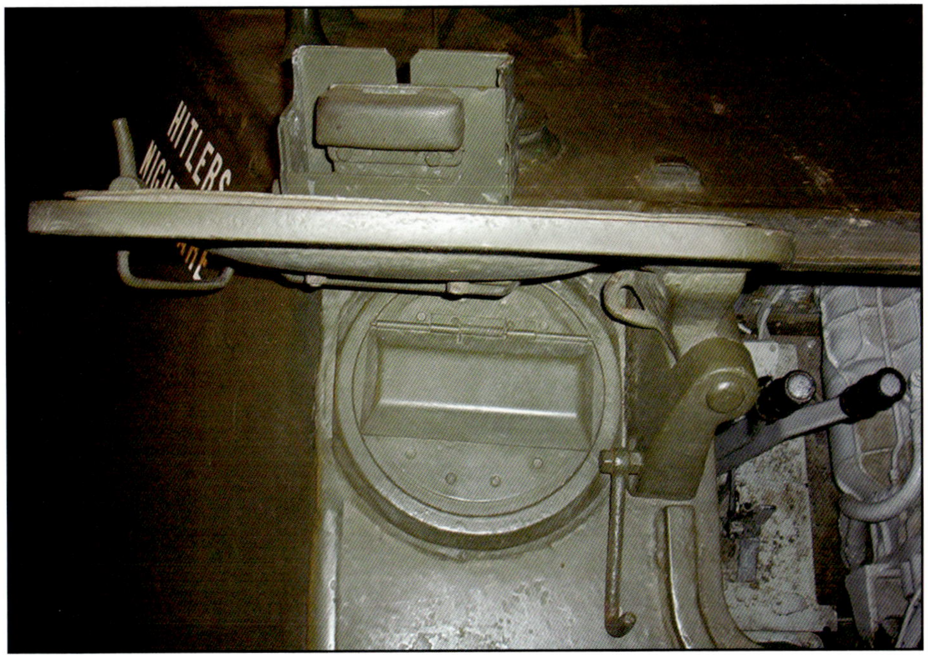

Top left: The driver's and co-driver's hatches are large steel castings with the large hinge plate welded to their outer edge. The hatches are equipped with a swiveling mount for a removable standard M6 glass block periscope. **Top right:** Here the co-driver's hatch is seen from the left side. The shape of the large hinge plates was designed to allow the hatches to open in the tight confines below the gun shield. The surrounding lip serves as a bullet splash. **Above left:** The driver's hatch opens and turns out and away from the turret. The inner mount for the periscope can be seen with its head pad. The left side of the transmission and the driver's brake handles are visible to the right. **Above right:** Outboard of the driver's hatch an additional swiveling periscope is mounted into the hull roof. The L-shaped rod attached to the hinge is a lock to keep the hatch plate safely in the open position while driving. (Don Moriarty, all)

Top left: The hatch plate has a lip just inside its circumference that keeps the hatch centered in the opening. The driver's lower seat pad and the gearshift lever are visible inside the hatch opening. Note the texture of the gun shield. **Top right:** The engine deck is similar in layout to the M4A2 Sherman. The engine deck doors are made up of welded pieces of steel rod and plate. The fillers for coolant, lubricating oil, and Diesel all have armored covers. **Above left:** The right forward portion of the deck has filler positions for the right side Diesel tank and engine lubricating oil. The fuel capacity of the M10 was 165 gallons, giving the tank destroyer a range of approximately 200 miles. (Don Moriarty, three) **Above right:** The left side deck has armored caps for the fill points of the left side tanks. Each filler is marked with a label to identify the proper fluid to be used. The box-like projections houses the fire extinguisher control handles.

Top left: The two central grill doors give access to the top of the two GMC Diesel engines. These doors allow the intake of cooling air that is directed through the radiators by a large fan mounted at the rear of each engine. **Top right:** The "U" shaped fitting at the rear edge of the deck is a gun travel crutch. The central filler is for coolant water for the radiators. The axe and track pry bar are stored in brackets on the rear deck. **Above left:** The M10 is armed with the 3-inch gun M7 that was developed from an antiaircraft weapon. The gun had superior armor penetration when compared to the 75mm M3 gun of the M4 Sherman. The barrel's rifling is visible in this shot. (Don Moriarty, three) **Above right:** The gun barrel of the M10 was extremely heavy, causing a severe imbalance of the turret. This made it very difficult to traverse the turret when the vehicle was on anything but level ground.

Top left: The gun shield at the turret front was a large casting with a maximum thickness of 2 $^{1}/_{4}$-inches, providing the most substantial protection of any section of the vehicle. Note the semi-circular detail on the shields left side. **Top right:** Lifting rings are provided on both outer edges of the shield to aid in removal and installation of the gun assembly. The central collar around the gun barrel is a separate casting welded into the shield assembly. (Don Moriarty, two) **Above left:** The opening for the gun sight is enclosed by a cover which the gunner pivots up and to the right to use the sight. The cast steel cover is marked with foundry numbers and is welded to its mounting pivot post. **Above right:** The casting marks for the gun shield is located at the center of its top edge. The part number is E6475 and the foundry mark denotes the manufacturer, National Roll & Foundry, Avonmore, Pennsylvania.

Top left: The mounting bosses for the proposed appliqué armor plates are also found on the turret sides. The armor thickness of the turret side plates is 1 inch, slightly thicker than the sides of the upper hull. (Don Moriarty) **Top right:** The M10 was plagued by its imbalanced turret throughout its service life. Various attempts were made to reduce the problem. This original style turret is equipped with the production counterweights. **Above left:** The counterweights weighed a total of 3,600 lbs. yet did not completely cure the problem of difficulty traversing the turret on a slope. The pole-like structure is a support for a tarp to cover the turret opening. **Above right:** The M10 was originally conceived as a lighter, more mobile, harder-hitting weapon than the M4. With the large turret counterweight the vehicle's weight surpassed 30 tons, roughly only 2 tons lighter than the Sherman.

Top left: Just ahead of the counterweights in the intersection of the rear turret plates is mounted the pintle for the .50-caliber antiaircraft machine gun. This was originally moved to this position to help the turret imbalance. **Top right:** Each counterweight is a large single piece casting. Smaller weights had been initially tried as a quick fix for the turret and were installed in the field. These smaller weights were made from steel, iron, or at times lead. **Above left:** The weights shown are the style designed for factory installation, and many were installed on earlier vehicles built without counterweights. It is common to see unit markings painted in the central portion of the weights. **Above right:** The forward portion of the turret is covered by a partial roof of ¾-inch armor plate. There are three of the folding legs for a canvas turret cover attached to the roof. A travel lock which attaches to the breech is at the center. (Don Moriarty, three)

Top left: At the rear of the main gun is the massive cast breech, missing the sliding breech block here. Forward of the breech are the recoil cylinders. The link from the gun to the roof is part of the elevating mechanism. **Top right:** The gunner in the M10 was placed on the left side of the gun, opposite of that in the M4. He was provided a monocular direct sight telescope M70G, which is missing its padded brow piece here. **Above left:** The mannequin occupies the gunner's position, with his traversing hand wheel to the front, folding seat behind, flanked by two turret locks. A second lock was added in production due to the locks breaking during travel. **Above right:** To the right of the gun are the seats for the commander at front and the loader to the rear. The seats were generally folded up during combat. The large tubular recoil guard folded down when needed. (Don Moriarty, all)

In a view of the left rear of an M10 turret, a 3-inch round is in the bottom of three cradle-type ready racks. To the right are two boxes for hand grenades, below which is the gunner's folded-up seat. Bolted to the turret ring are two turret locks, which were disengaged before putting the gun into firing order. (Marc Sehring)

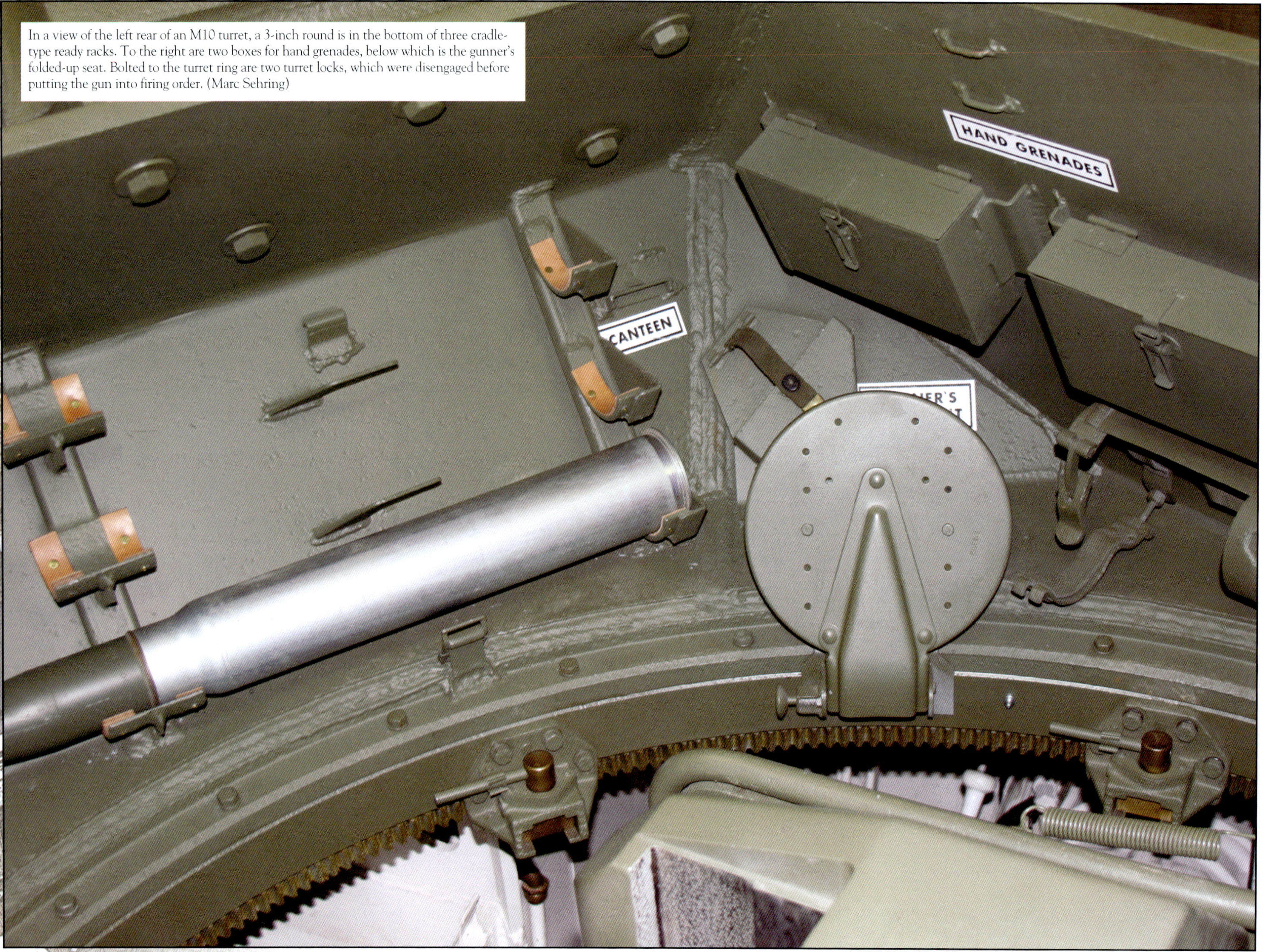

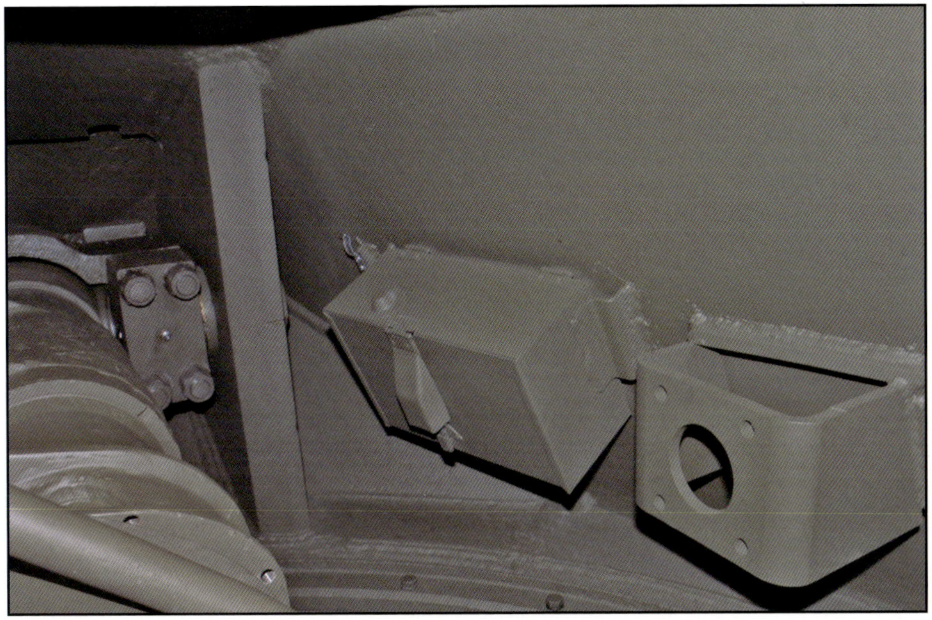

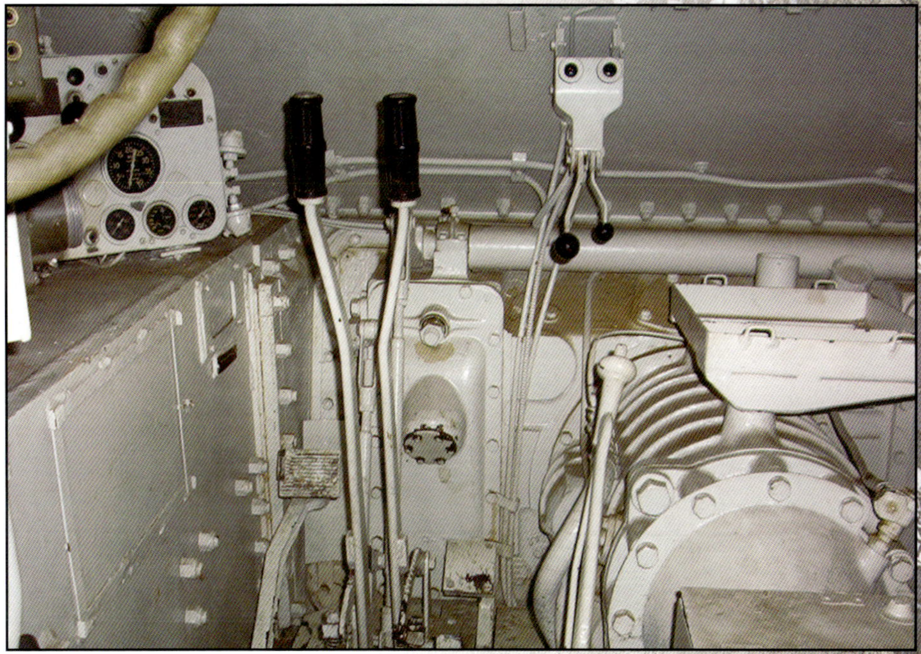

Top left: From front to rear on the right side turret interior wall are mounted a box for a spare periscope and a mounting bracket for the panoramic artillery sight for the commander. (Don Moriarty) **Top right:** The turret of the M10 was only equipped with a manual traverse, which was quite slow. The traversing wheel and gear assembly was mounted on the turret ring at the left front in the gunner's area. **Above left:** Immediately forward of the gunner's position is the driver's seat and controls. The M10 was based on the chassis and running gear of the Diesel powered M4A2 Sherman and shared common controls and instruments. (Don Moriarty) **Above right:** From left to right the driver's controls include the clutch pedal, left and right brake control handles used to steer the tank, accelerator pedal, and gear shift lever. The two handles above are lockouts for the 2 clutches. (Don Moriarty)

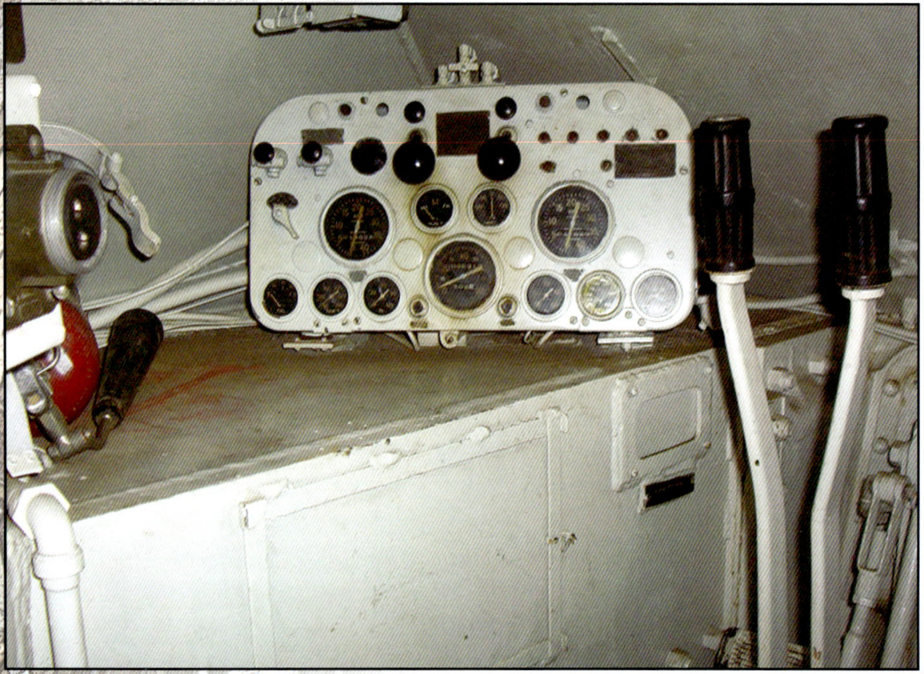

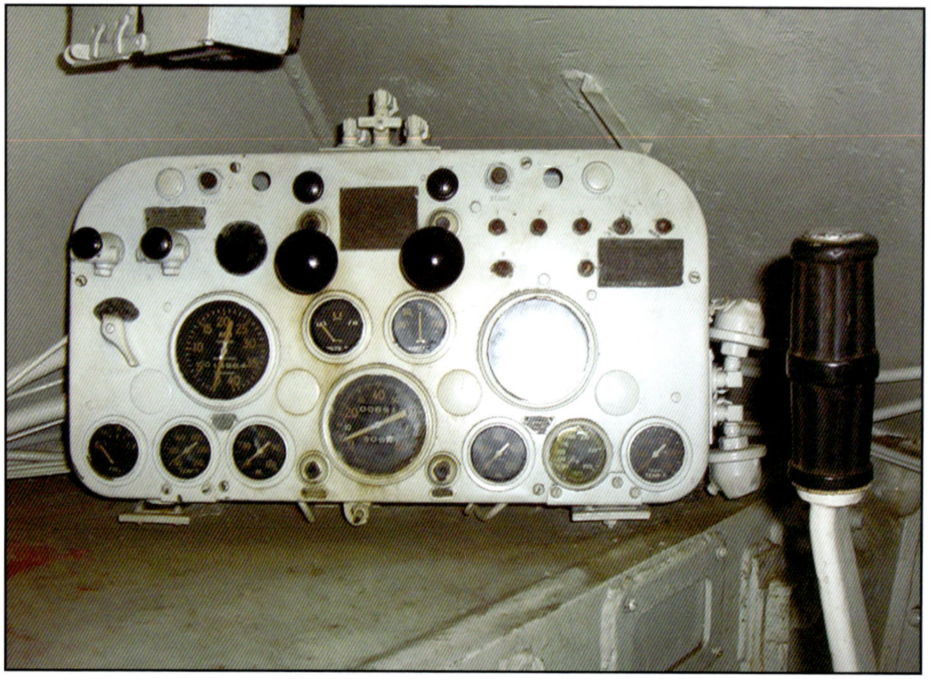

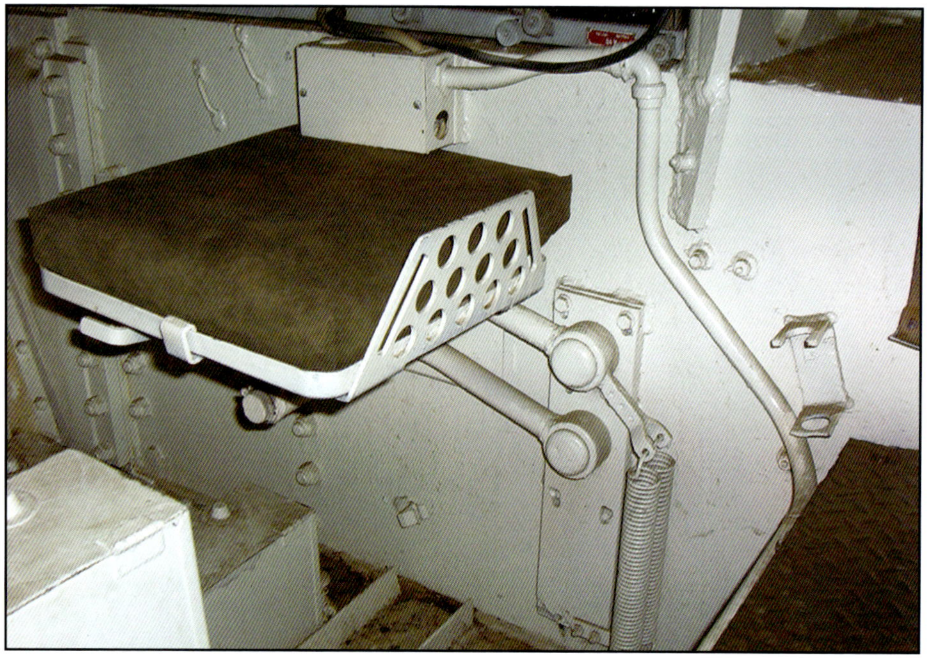

Top left: The instrument panel was equipped with controls and gauges to monitor the two engines separately. The rectangular framework on the hull sidewall was for mounting the maintenance lubrication chart for the vehicle. **Top right:** The tachometers are located at the top with amperage and voltage between them. Below are the fuel gauges, two engine coolant temperature gauges, two oil pressure gauges, transmission temperature gauge and a speedometer. **Above left:** Centrally located between the driver and co-driver is the transmission. It was a five-speed synchronized unit with a lockout button on the shifter for selecting first gear and reverse. The rack above is for spare periscope heads. **Above right:** The driver and co-driver's seats were designed to be used at two different levels; lowered for operating with the hatches closed and raised to operate with heads out of the open hatches. (Don Moriarty, all)

Top left: The right front sponson area houses the vehicle's radio. The area was large enough to accommodate the SCR-528, as seen here. Many vehicles were converted to this larger unit with a transmitter and two receivers. **Top right:** Behind the co-driver's seat in the hull floor plate is located a crew escape hatch. Pulling back on the central handle releases the 3 catches that hold the heavy hatch plate, allowing it to drop to the ground below the vehicle. **Above left:** Ammunition was stowed in the sponsons in four simple racks. The rounds were stacked in the racks inside their fiberboard packing tubes and held in place by web straps. These racks held forty-eight rounds. **Above right:** The box-like projections on each side of the rear fighting compartment cover the generators for each of the Diesel engines that project through the firewall. The electrical system was 24 volts with a pair of 12-volt batteries in series. (Don Moriarty, all)

On the final drive assembly of this M10 is a step made of welded steel strapping material. A grab handle to assist the crew in mounting and dismounting the vehicle is secured to two of the bolts for supplemental armor on the glacis. Tow hooks are installed on the lower part of the final drive assembly. (Marc Sehring)

The late-production M10s were built with a redesigned turret which had nearly vertical rear walls and a lighter, longer counterweight. These are often called “Duckbill Turrets” due to the counterweight shape.

Numerous footman loops were welded to the exterior walls of the turret. The crew's personal kit is often seen strapped to these loops with lengths of webbing. The interior of the vehicle had little space for stowing gear.

Top left: The shape of the side armor plates of the turret was changed at the upper rear corner where they meet the rear walls. This change also made the interior of the turret quite a bit roomier for the crew. **Top right:** The new turret initially retained the circular mounting bosses and bolts for the proposed appliqué armor kits that were never produced. The later production hulls and turrets eliminated these bosses from the sides. **Above left:** The new shape of the counterweights is evident from the rear. The new counterweight was over 1,000 pounds lighter and yet better balanced the turret. This reduced the effort required to traverse the turret while on a slope. **Above right:** The vertical rear walls greatly increased the available area for the loader to operate to the right and rear of the breech. The rectangular plate at the lower left is a mount for the commander's binocular case.

The commander was also provided an elevating hand wheel on the right side of the gun. The operating handle for the breech is on the right side for the loader. Note the coil spring attached to the folding recoil guard.

The redesigned rear turret walls reduced the number of plates and welds. The ammunition ready racks remained in the same place, while the rear turret lift ring was repositioned to the interior below the M.G. mount.

The British Achilles

Although the British and Commonwealth forces, as well as certain Allies such as the Poles, were supplied with M10s with 3-inch guns, the British also developed a version of the M10 armed with the very potent 17-pounder quick-firing antitank gun. In this photo is a column of these vehicles, designated both the M10C and the M10 17pdr. The nearest one has what appears to be a supplemental shield with a vision space below it, above the front part of the turret.

The 17-pounder gun of a British Achilles is engaging enemy pillboxes and an 88mm gun on the western frontier of Germany in or around October 1944. The battery to which this vehicle was assigned already had racked up 21 German tanks killed. Two wooden crates are lashed to the rear of the hull. (NARA)

A British Achilles nicknamed "Glengarry II" was photographed on the Continent in or around September 1944. This vehicle served with G Troop, 46th Battery, 91st Antitank Regiment, Argyll and Sutherland Highlanders, British Eighth Corps. The crew welded a horseshoe to the final-drive assembly for good luck. (NARA)

The German soldier on the left is pointing to two tightly spaced penetrations on the side of the turret of an Achilles in Northern France in summer 1944. The vehicle carries census number S23781. Another penetration of similar gauge is on the side of the rotor shield, and there may be another one on the hull above the head of the man to the right. Other damage includes a broken antenna mount and destroyed grouser rack, which suggests another shell probably struck the hull in that area. (BA 299-1818-05)

The breech, recoil guard, and spent-casing deflector of the 17-pounder in an M10C turret is viewed from the front, with ready rounds stored on the rear of the turret. The weapon used in the M10C was, formally, the 17-pounder Mk. V in Mounting No. 3 Mk. 1. Round, folding seats for crewmen are along the turret ring.

The M10 was originally designed to accept the US 105mm Howitzer and the British 17pdr gun in its gun mount. The British saw the conversion of the M10 as an expedient way to get the formidable 17pdr onto the battlefield on a mobile mount. The Royal Ordnance Factories performed the conversions. The Factories were in Woolwich (London), Nottingham, Radcliffe, Ellesmere Port and Glasgow. The British used the weapon more like a self-propelled defensive antitank gun than in the tank destroyer role. The Achilles was also issued to other Commonwealth armies, and served in WWII with Canadian, Polish, and New Zealand units. (John Blackman)

By April 1945 1,017 of these conversions had been made, the majority being carried out on the later turret design with the Duckbill counterweight and straight rear walls. Although today popularly known as the Achilles, during the war the correct title was 17pdr M10 SP Mark IC or IIC. The "C" denoting the 17pdr gun. Internal ammunition stowage was altered, reducing the total load of rounds to 50. The radios were replaced with British No. 19 units. (John Blackman)

The Ordnance Q.F. 17pdr was the most powerful Allied antitank weapon when introduced in 1942. The British had rightfully responded to the escalating quality of both tank arms and armor that was being driven by the tank battles on the Eastern front. The gun was able to penetrate as much as 2 inches more armor at comparable ranges when compared to the standard 3-inch gun of the M10. (John Blackman)

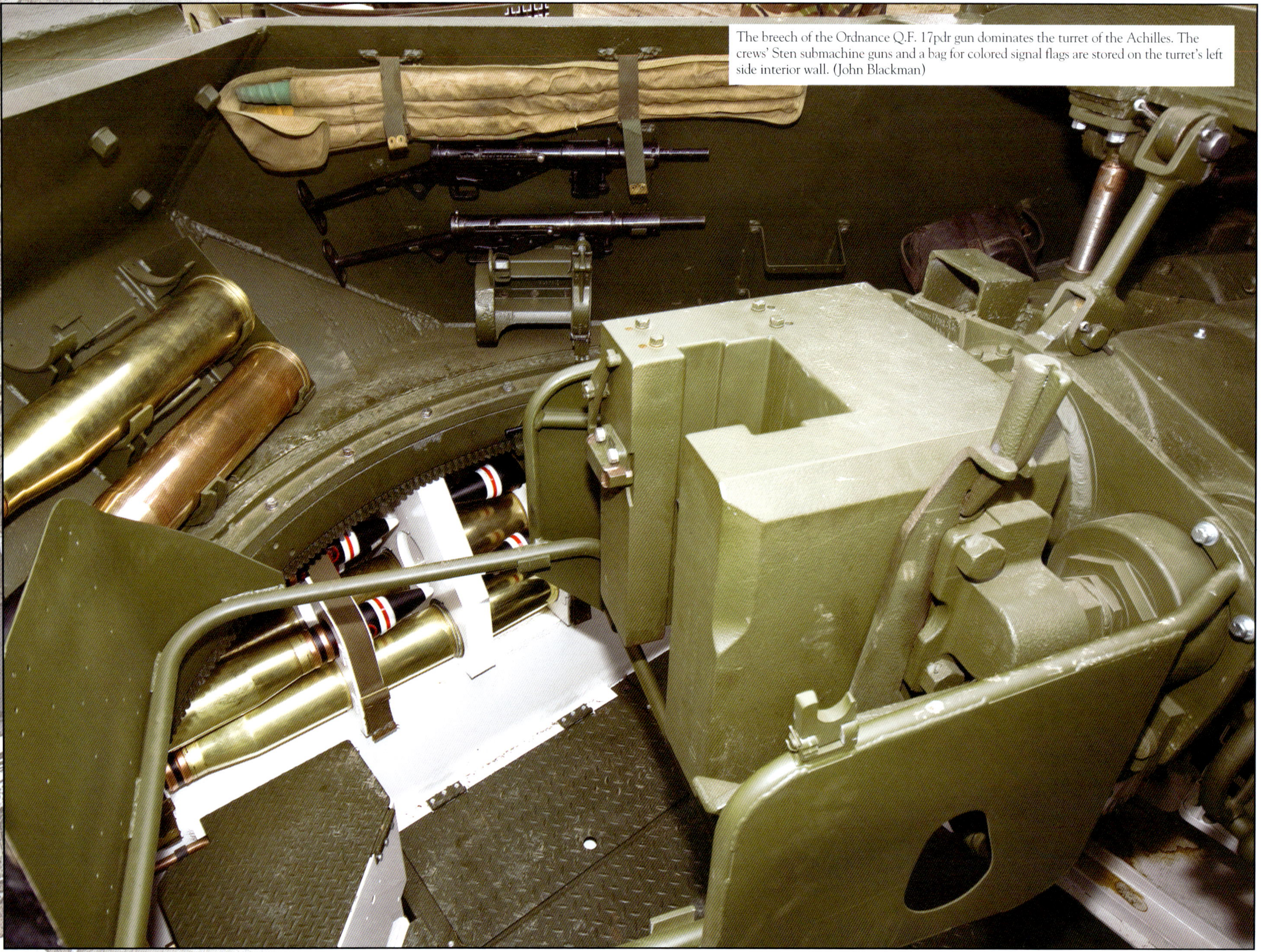

The breech of the Ordnance Q.F. 17pdr gun dominates the turret of the Achilles. The crews' Sten submachine guns and a bag for colored signal flags are stored on the turret's left side interior wall. (John Blackman)

Part of the Achilles conversion was making provisions for storage of the new ammunition. The sponson racks stowed the rounds, which had a larger case than the 3-inch rounds, without any covering in sponson racks. (John Blackman)

Left: Much like the standard M10, the Achilles recoil guard was designed to fold up and out of the way when not needed. This provided much more space for work or movement in the area behind the gun.

Right: The gunner's position on the 17pdr gun is on the left side which is the same as on the unmodified M10. This made the conversion much more straightforward than on the similarly armed 17pdr Sherman Firefly. (John Blackman, both)

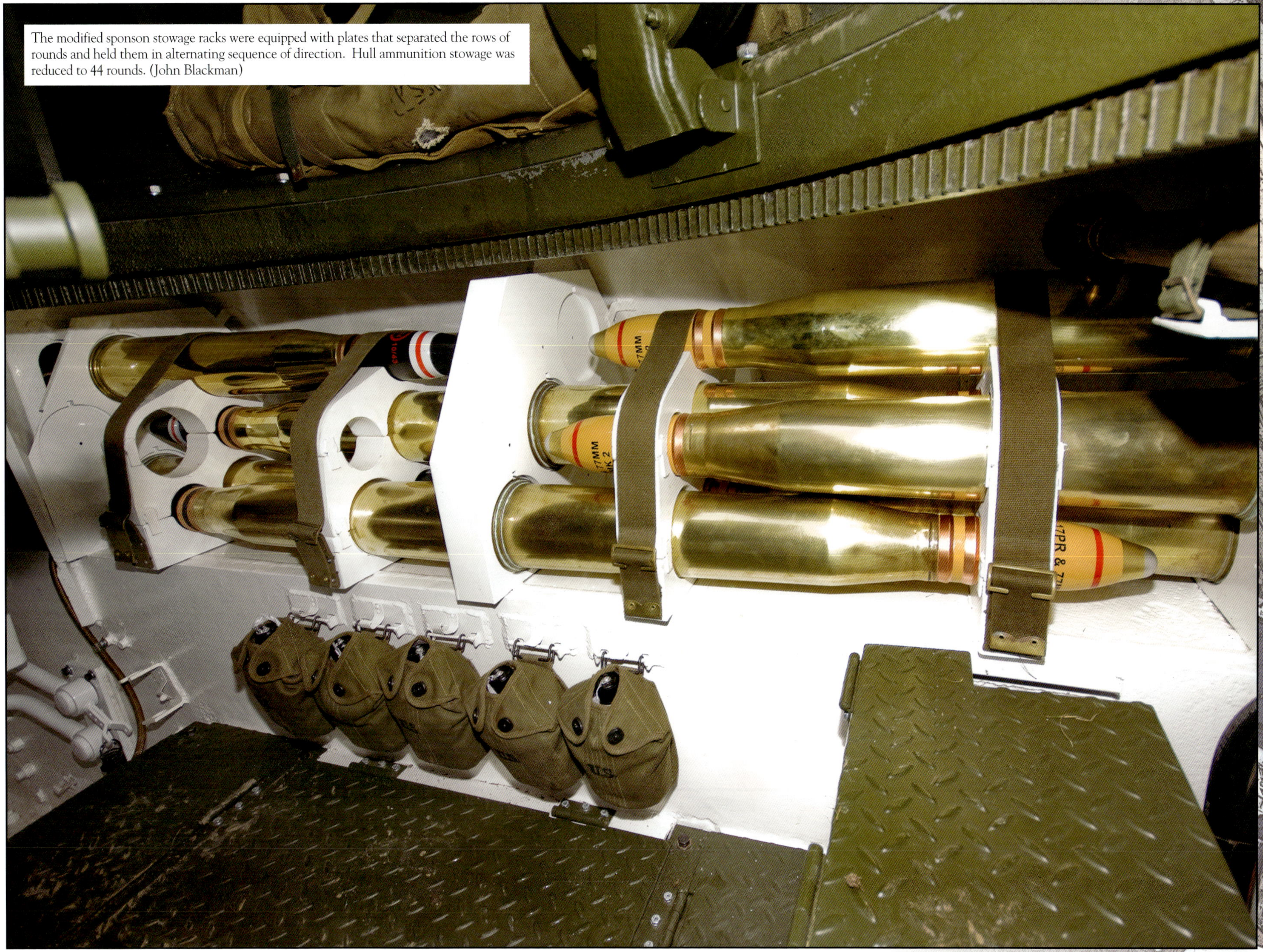

The modified sponson stowage racks were equipped with plates that separated the rows of rounds and held them in alternating sequence of direction. Hull ammunition stowage was reduced to 44 rounds. (John Blackman)

Stowage of ready rounds in the turret remained at 6 rounds, although the racks were modified. The toothed gear of the turret ring is visible which meshed with the traversing gear to manually rotate the turret. (John Blackman)

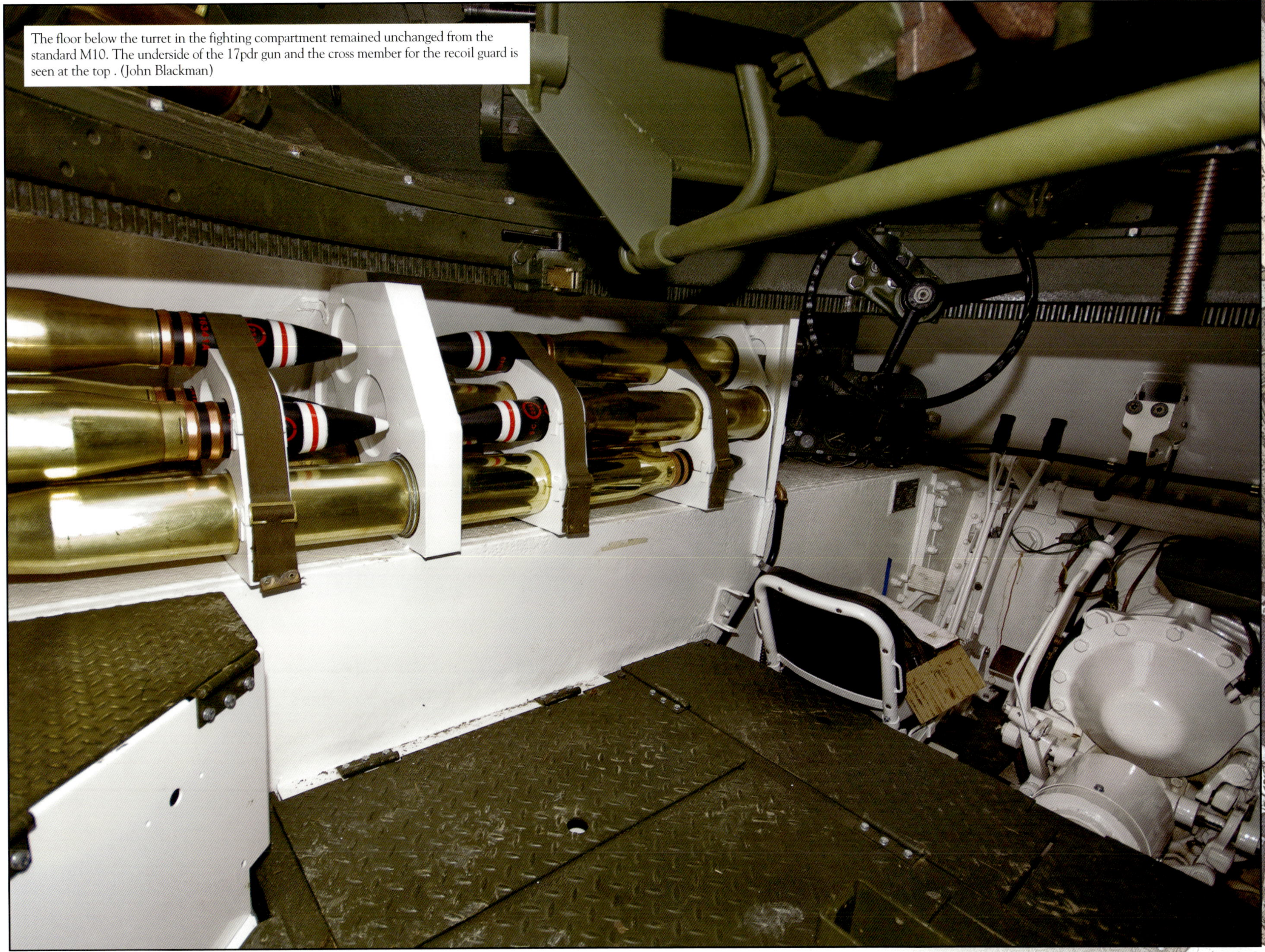

The floor below the turret in the fighting compartment remained unchanged from the standard M10. The underside of the 17pdr gun and the cross member for the recoil guard is seen at the top . (John Blackman)

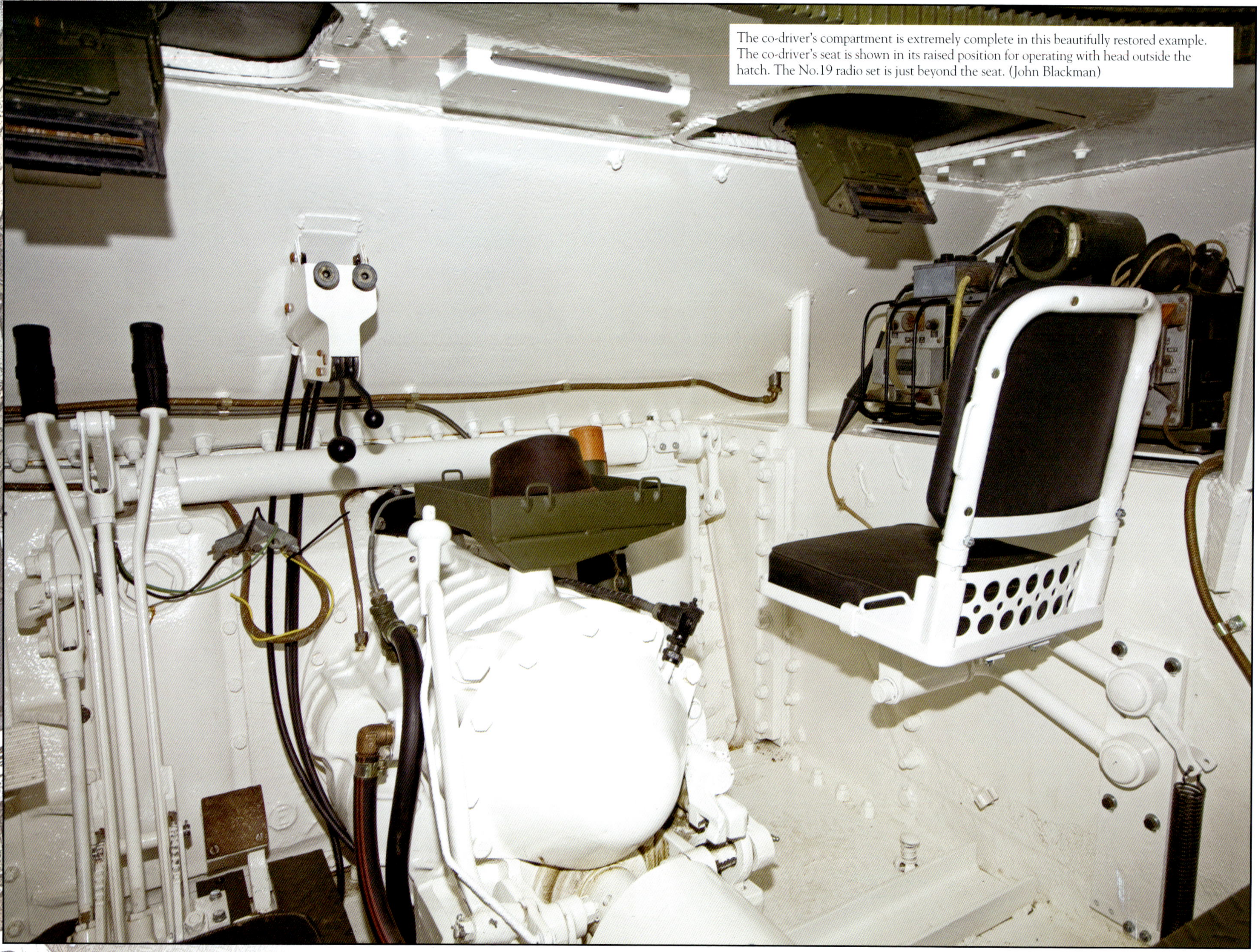

The co-driver's compartment is extremely complete in this beautifully restored example. The co-driver's seat is shown in its raised position for operating with head outside the hatch. The No.19 radio set is just beyond the seat. (John Blackman)

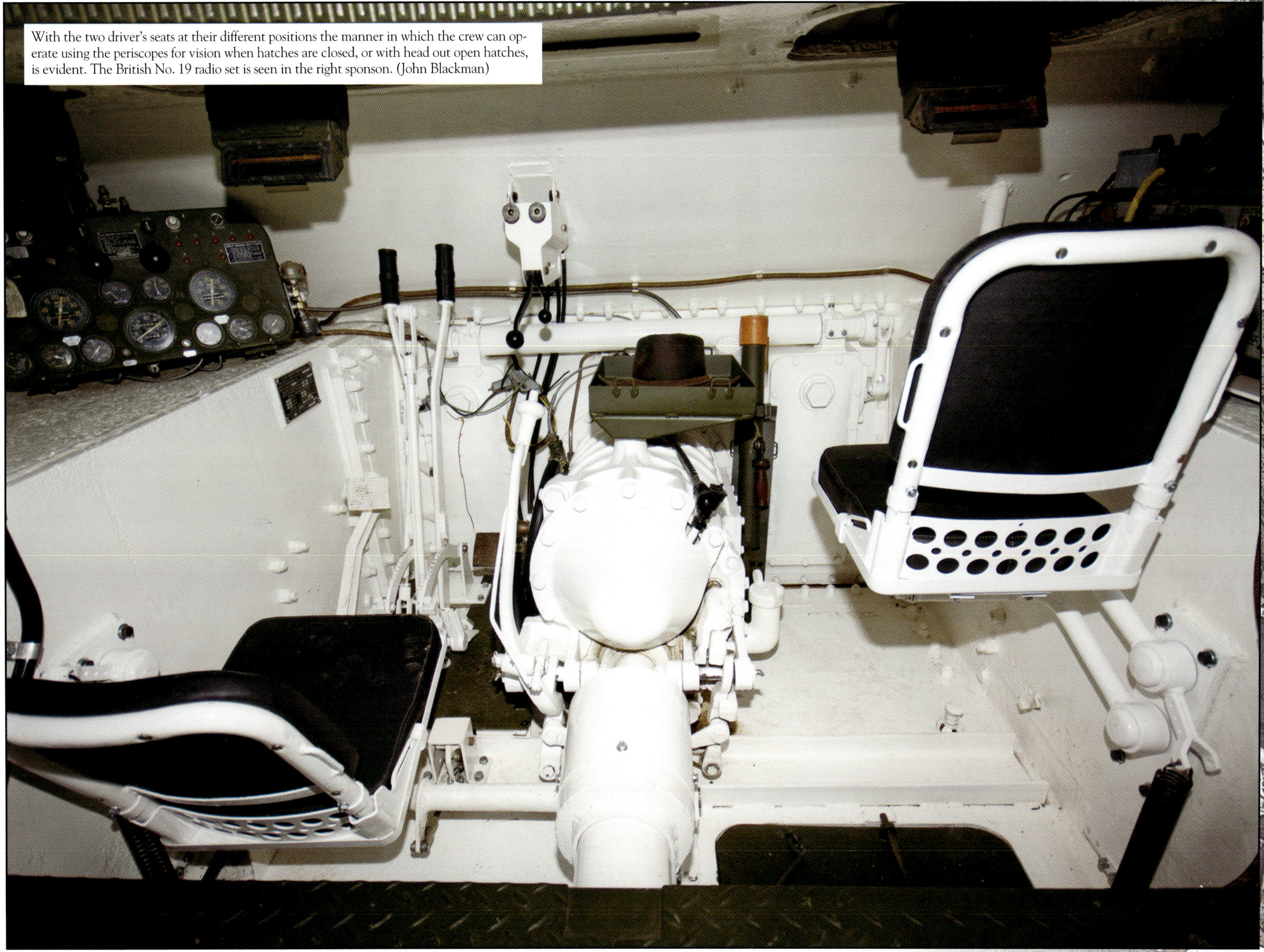

With the two driver's seats at their different positions the manner in which the crew can operate using the periscopes for vision when hatches are closed, or with head out open hatches, is evident. The British No. 19 radio set is seen in the right sponson. (John Blackman)

Left: Looking across the transmission assembly into the driver's station in the left front of the vehicle, this area is unmodified from that of a standard M10. The driver's rotating periscopes are visible at the top center.

Right: This beautifully restored vehicle carries two spare wheels on the differential cover. The T48 rubber chevron tracks are heavily worn. The siren is missing from the bracket on the left glacis plate. (John Blackman, both)

The Achilles IC is equipped her with the solid road wheels with the stamped metal faces and a welded idler wheel, identified by the open slots. The fuel can rack ahead of the grouser stowage is a post-war addition. (John Blackman)

A sheet metal tray is located below the louvered engine deck doors above the engine. The tray serves both to keep water out of the area at the top of the engines and to direct cooling air rearward to the fans and radiators. (John Blackman)

The tray or guard is hinged and swings upwards to allow access to the engine for service. Fuel filters are mounted on the forward wall of the engine compartment. The three cylindrical fittings on each side are air filters. (John Blackman)

The Achilles required the addition of a cast armored collar to reduce the opening in the gun shield due to the smaller diameter of the 17-pounder barrel. This Achilles I has a full set of solid spoked road wheels. (John Blackman)

The late production M10s re-armed with the 17 pdr. were named Achilles IIC. The later production turret was far more prevalent. The turret rear area with the straight walls provided more room for loading the weapon. Like many later production M10s, it mounts the E8543 FDA. The example shown has some post-war additions such as the rear view mirrors, siren guard, fuel can racks, and antennas for a later pattern radio, but is in excellent condition and is very complete. The tent-like canvas cover for the open top turret is shown erected. (John Blackman)

Another restored Achilles IIC churns up dust while on display at the Tank Museum's Tank Fest in Bovington, England. This vehicle is authentically equipped with personal gear for the crew, fuel cans, canvas water buckets, camouflage netting, and even a spare set of drive sprockets. The muzzle is protected by a canvas cover. (John Blackman)

Top left: The 17pdr barrel was equipped with a muzzle brake which helped reduce the gun's recoil and directed muzzle blast to the sides. This reduced the amount of debris kicked up when firing the gun. **Top right:** The counterweight just behind the muzzle brake was necessary to balance the gun mount on its trunnions. This clamshell style fitting is fixed in place with bolts through the casting. **Above left:** The barrel of the 17 pdr. gun was smaller in diameter than the 3-inch gun of the M10. The installation required the addition of a cast armored collar to reduce the opening in the gun shield. **Above right:** The gun shield was further modified by increasing the small round opening for the gun sight to a larger vertical slot in the left side. This gave the gunner a direct view out of the turret above the gun sight.

A close-up of the E4186 differential and final drive housing. Note that the armor on the bulges over the final drives (near drive sprockets) is flush with the edge of final drive covers. At some point, inside towing lugs on the final drive have been added and then later cut away. (Chris Hughes)

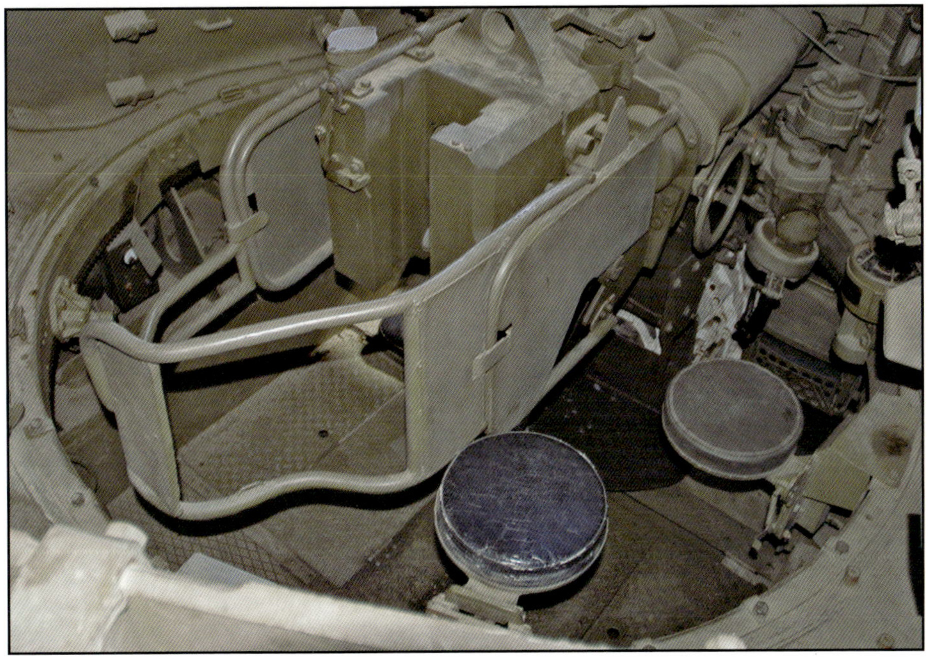

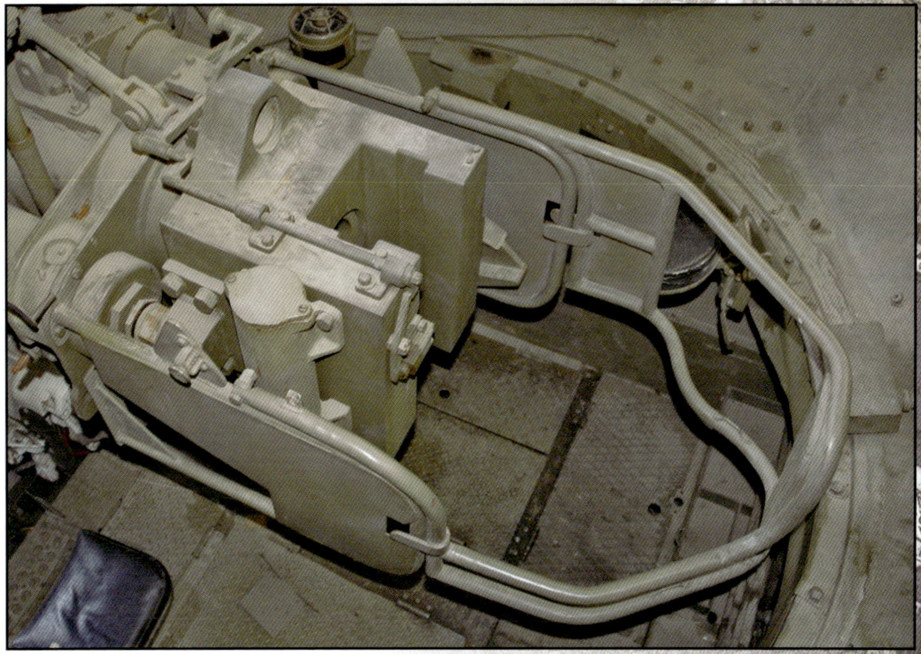

Top left: The differential cover caries a casting mark in its upper right surface. The drawing number of the casting is E4186. The U in the keystone shape is the mark of Union Steel Castings Pittsburg, Pennsylvania. **Top right:** The majority of M10s and M10A1s converted to Achilles were built with the later turret. The vertical rear walls of the turret gave much more working space for the crew to handle the long rounds of the new gun. **Above left:** The loader and commander's seats are folded down on the right side of the gun. This Achilles has been further modified with a hydraulic power traverse unit similar to that of the M4 Sherman, and an azimuth indicator. **Above right:** The left side of the breech assembly with the breech spring is seen enclosed inside the vertical tubular fitting. Just ahead of the spring is the back end of the left side hydraulic recoil cylinder. (Chris Hughes, all)

An excellent view of the details of the breech assembly of the 17pdr gun is afforded with the recoil guard removed. The area in the forward right of the turret differs from the example shown with the power traverse. (John Blackman)

Top left: The fact that the M10 was designed with a gun mount that would accept both the 105mm howitzer or 17pdr gun made the Achilles conversion a much simpler job. Elevation range was reduced slightly with the new gun. **Top right:** The Achilles was supplied to the anti-tank units of British armored divisions. It was used in the role of a mobile anti-tank gun rather than as a tank destroyer. **Above left:** The ammunition racks missing from the right rear turret wall provide a good view of the mounting straps and bolts for the turret counter weight. The pintle mount for the machine gun is at the top center. **Above right:** The turret ready racks held 6 rounds of 17-pounder ammunition. The loader's seat is folded down in the turret right rear; to the left above the geared ring is one of the turret locks. (Chris Hughes, all)

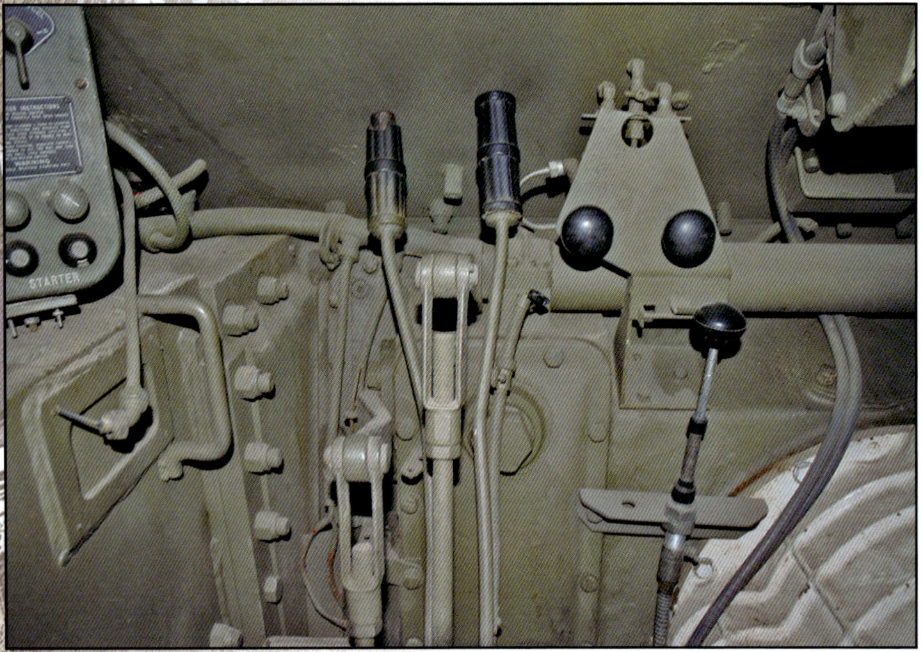

Top left: The design of the "duckbill" counterweight and the change of the turret made it possible to balance the turret with a lighter weight. The openings in the castings gave the crew additional space for storing gear. **Top right:** The transmission on this Achilles IIC is fitted with a generator in the power take off at the top of the gear box. This is a common fitting on some Sherman based variants. **Above left:** The driver's controls are the same as the standard M10. Parallel to the brake controls handles are adjustable links to the brakes. The removable panel forward of the controls is for access to the brake linings. **Above right:** Located at the base of the brake control handles are the clutch and accelerator pedals. Note the ratchet assemblies at the base of the brake handles for placing them in a set position. This also serves as a parking brake. (Chris Hughes, all)

Top left: This late-production vehicle has a different instrument panel layout for monitoring the twin diesels. Compare this to the panel on page 88. Behind the panel are an electrical junction box and the main control box. **Top right:** This assembly is the clutch lockout box. It was used to disengage one of the two twin Diesel engines. **Above left:** The fighting compartment floor is made up of folding panels of tread plate. This panel is for access to the twin batteries, the hole in the panel. **Above right:** Atop the transmission between the driver's positions is a pair of boxes for stowing spare M6 periscope heads. The M6 was the standard armored glass periscope for all U.S. armored vehicles from 1942 to 1945. (Chris Hughes, all)

The right side suspension is fitted with three styles of solid road wheels. The rear wheel is fairly common on late war M4 based vehicles and has a simplified smooth stamped face when compared to the spoke wheel. Note the wider hubs on these wheels, which contained high capacity bearings for greater load strength. The front wheel on the rear bogie is a solid stamped type that is often used as a post-war replacement. The drive sprocket plate is the solid flame-cut design. The bogies are equipped with the straight return roller brackets. The end connectors are the large profile post-war replacement type. (John Blackman)

The canvas cover for the turret roof opening was equipped with handles at the points where the grommets fit over the legs. This aided the crew in positioning the hole over the prop, especially if the cover was wet. (John Blackman)

An M10A1 from a very late production batch which was built without the circular mounting bosses from the turret side armor plates was used for the conversion of this Achilles IIC. This is a beautifully preserved example with the spare track racks on the upper portion of the differential cover intact. The only postwar modifications visible are the pair of antenna mounts on the right glacis. The presence of the E8543 FDA indicates late 1943 or early 1944 production. (John Blackman)